SCIENCE ON THE EDGE

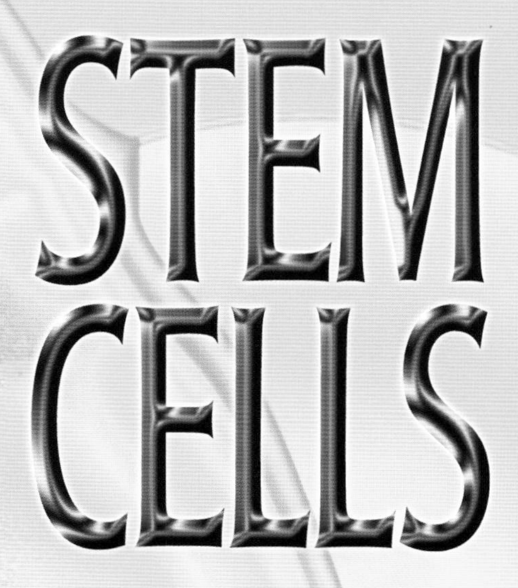

WRITTEN BY

JENNY TESAR

THOMSON
GALE

San Diego • Detroit • New York • San Francisco • Cleveland • New Haven, Conn. • Waterville, Maine • London • Munich

For more information, contact
The Gale Group, Inc.
27500 Drake Rd.
Farmington Hills, MI 48331-3535
Or you can visit our Internet site at http://www.gale.com

Photo credits: cover © Greenhaven Archives; pages 4, 6, 7, 8, 9, 10, 11, 12, 14, 15, 16, 17, 18, 19, 20, 27, 28, 29, 30, 31, 32, 34, 40, 41, 43, 44 © CORBIS; pages 5, 6, 13, 21, 22, 23, 24, 25, 28, 36, 37, 38, 39, 42 © PhotoResearchers

LIBRARY OF CONGRESS CATALOGING-IN-PUBLICATION DATA

Library of Congress Cataloging-in-Publication Data

Tesar, Jenny E.
Stem cells / by Jenny Tesar.
p. cm. — (Science on the edge series)
Summary: Examines the very promising but controversial use of human stemcells in treating medical conditions ranging from burned skin to damagedspinal cords to various diseases.
Includes index.
ISBN 1-56711-787-2
1. Stem cells—Juvenile literature. 2. Embryonic stem cells—Juvenile literature. [1. Stem cells. 2. Cells. 3. Medicine.] I. Title. II. Series.

QH587.T47 2003
616'.02774—dc21 2002012416

Printed in China
10 9 8 7 6 5 4 3 2 1

TABLE OF CONTENTS

INTRODUCTION

You began as a single, microscopic cell. By the time you were born, you contained about 20 billion cells. Now you are several feet tall and your body consists of trillions of cells. The cells come in many different shapes and sizes. They include blood cells, skin cells, heart cells, muscle cells, eye cells, brain cells, and others. Even within these kinds of cells there is variety. For example, there are five different kinds of white blood cells.

How does a single cell become trillions of cells of different kinds? The single cell soon divides, and divides again and again, to form an embryo. Within several days, the embryo contains a cluster of cells that scientists called embryonic stem cells. These stem cells give rise to all the specialized cells in the body.

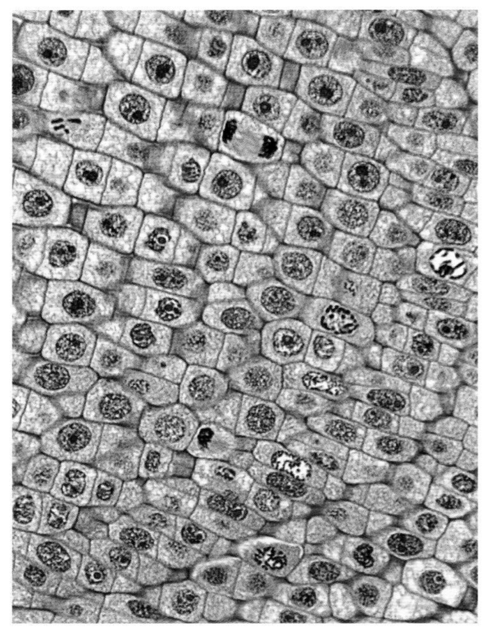

Microscopic view of human skin cells

Embryonic stem cells are found only in embryos. Children and adults, though, have another type of stem cell, called adult stem cells. These cells can differentiate into specific kinds of cells that are lost through normal wear and tear or during injury or disease. For example, adult stem cells in the skin develop into different kinds of skin cells.

Scientists are only in the early stages of studying and

This electron micrograph shows the structure of embryonic human stem cells.

Karen Hughes, counselor to President George W. Bush, was at the center of the public debate about funding stem cell research.

understanding stem cells, but they are excited about their discoveries and by the potential benefits of their research. Stem cell research is very controversial, however. It is the cause of much debate among politicians, religious leaders, and the general public. At the center of the debate are two conflicting viewpoints.

Some people are opposed to research on embryonic stem cells. Typically, an embryo must be destroyed to obtain embryonic stem cells. Some people believe embryos are human beings. They consider it immoral to destroy embryos for research or any other purpose. They want scientists to limit their research to adult stem cells.

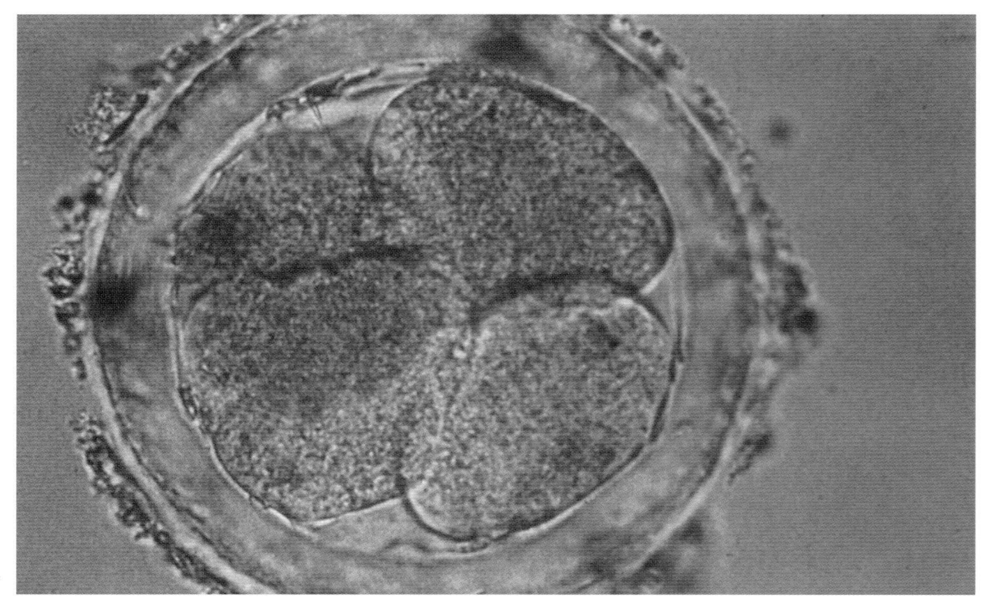
Microscopic view of a human embryo

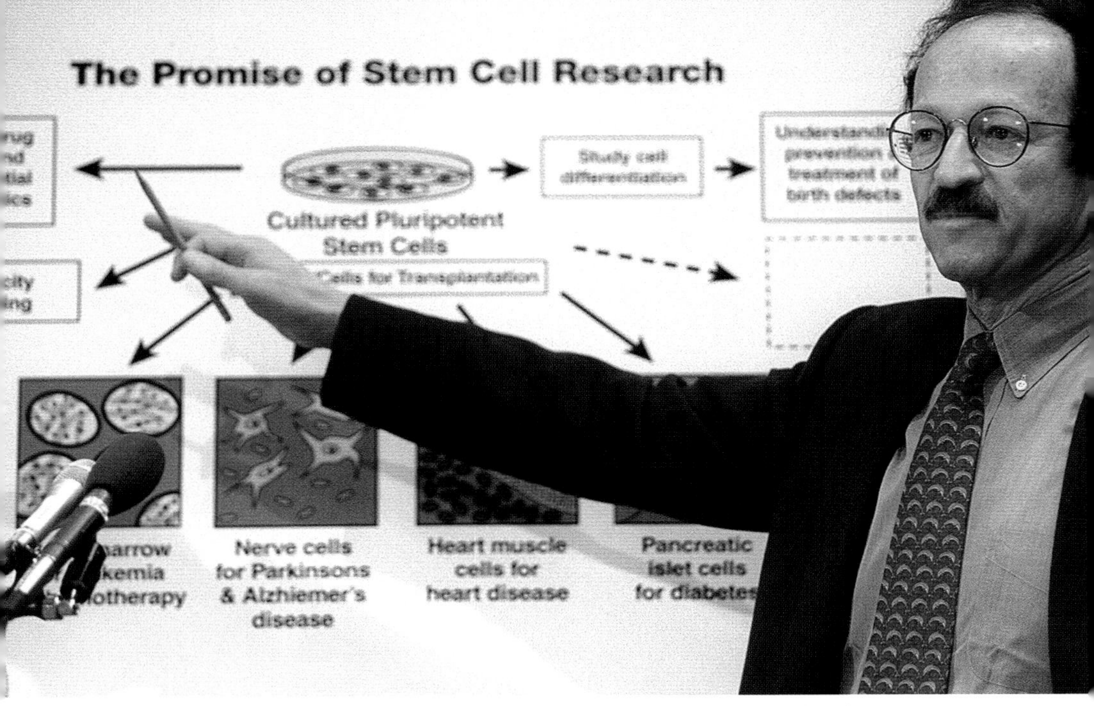

Dr. Harold Varmus, director of the National Institutes of Health, is a vocal supporter of stem cell research.

Supporters of stem cell research do not equate a four- or five-day-old embryo grown in a laboratory with a human being. They point out that embryonic and adult stem cells differ from one another, and thus they believe that it is important to study both kinds of cells. They focus on the exciting possibilities that may result from research: repairing or replacing damaged body parts, curing diabetes and other diseases, and developing more effective medicines.

These possibilities are still years away from becoming reality—if, indeed, they become reality at all. First, scientists must answer numerous questions, from how stem cells produce specialized cells to how they can be used safely and effectively in medicine. Obstacles, such as a body's tendency to reject foreign cells and organs, must be overcome. New techniques and technologies must be developed to isolate, purify, and grow stem cells in large quantities. All this helps make stem cells one of today's hottest areas of research among biologists and medical scientists.

CHAPTER 1

STEM CELLS AND THEIR DISCOVERY

One scientist who works with stem cells calls them "infant cells that have not yet chosen a profession." Stem cells can be described not by how they look but by what they can do. First, they are unspecialized. They cannot grow hair or digest food or send nerve signals. Second, they have the ability to "choose a profession." That is, they can develop into various kinds of cells. For example, human stem cells can form more than 200 types of specialized cells, including skin cells, blood cells, bone cells, fat cells, nerve cells, and muscle cells. Third, through cell division, stem cells can renew themselves for long periods of time. That is, they can make exact copies of themselves over and over again, until something triggers them to differentiate.

An embryonic stem cell like this one can reproduce itself many times until something triggers it to differentiate.

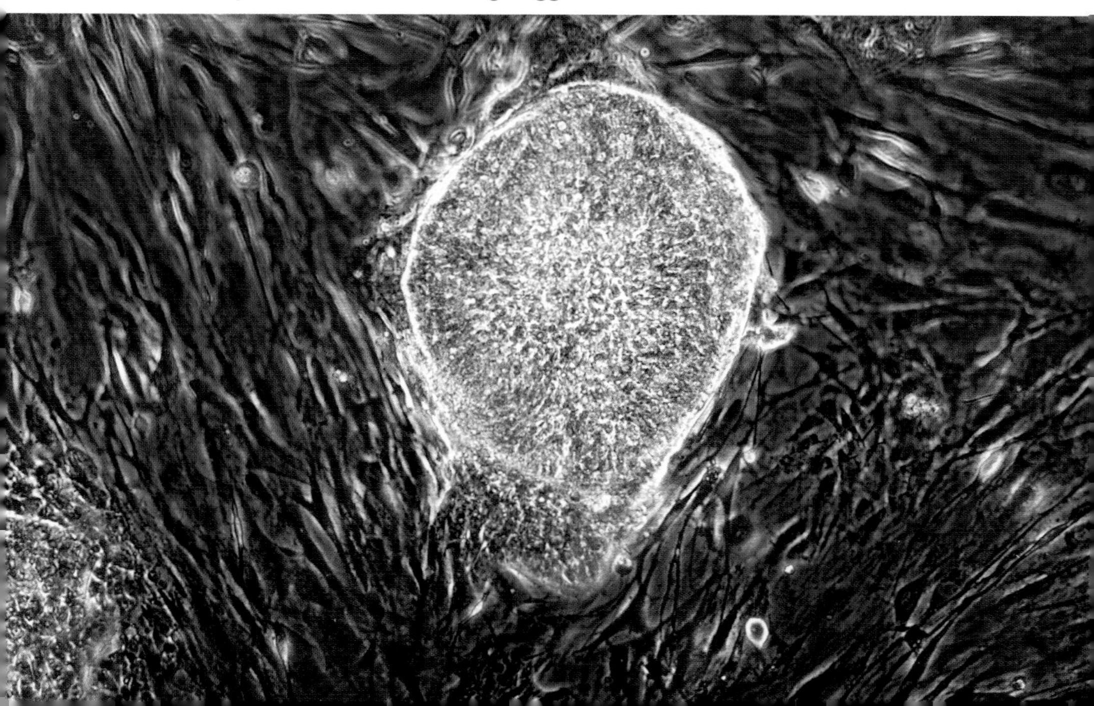

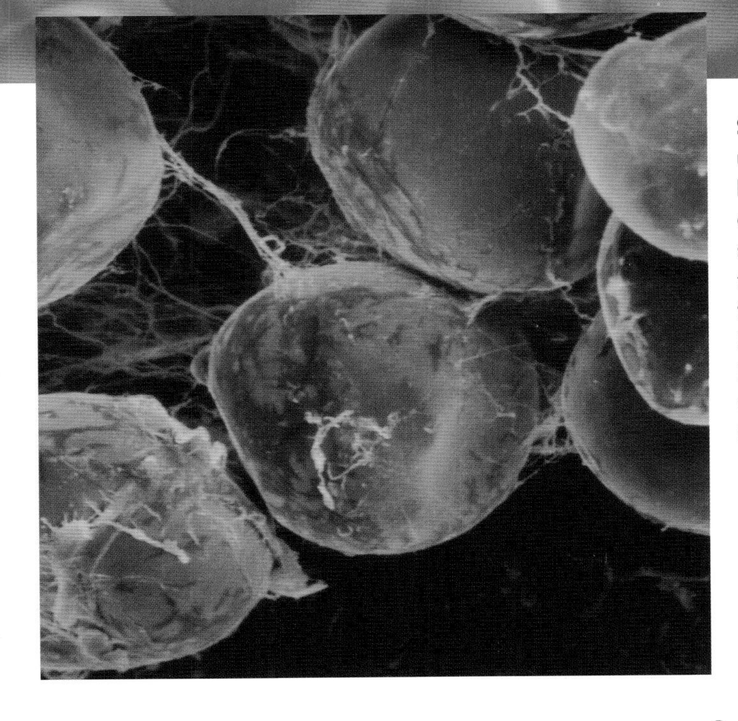

Stem cells are unspecialized before they differentiate, or specialize. Once they do, they can form more than 200 types of cells, including blood cells (above) and fat cells (left).

The existence of stem cells was first suspected early in the 20th century, as scientists studied the development of embryos. (An embryo is the first stage of an organism, before it is born.) The scientists theorized that an embryo contains powerful cells that act as seeds for the formation of all other types of cells. Indeed, these cells were called seed cells at the time. Scientists also suspected that seed cells are able to form certain, specific kinds of cells that exist in the body after birth and throughout life. In particular, because scientists knew that blood cells have a short life span, they believed that there were seed cells in the bone marrow that could form new blood cells.

Bone marrow played an important role in early stem cell research.

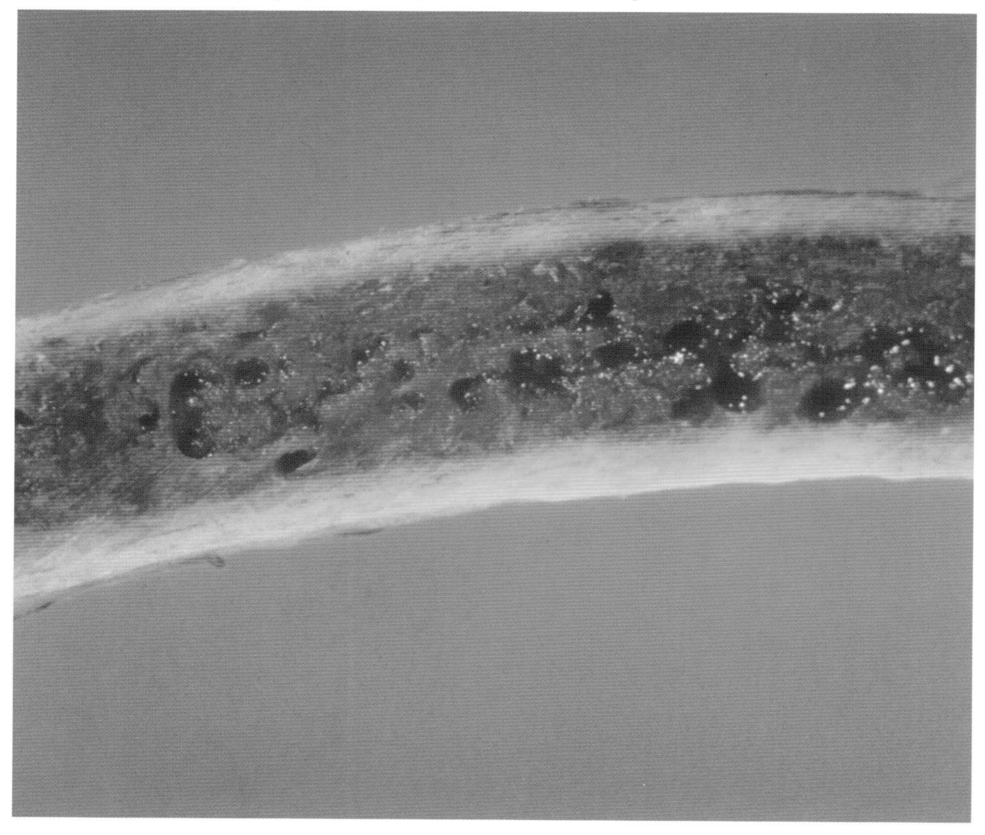

THE FIRST EVIDENCE

The first confirmation that stem cells exist came in the early 1960s. James Till and Ernest McCulloch at the Ontario Cancer Institute in Toronto, Canada, were studying how radiation destroys the blood cells of laboratory mice. They found that they could rebuild the mice's supply of blood cells and prevent the mice from dying by giving them injections of bone marrow cells from other, genetically similar mice.

Less than two weeks after they injected the bone marrow cells, Till and McCulloch examined the bone marrow and spleen of the recipient mice. They found lumps of blood-cell colonies. They expected to find only one kind of blood cell in each lump. Instead, they found that each lump contained all types of blood cells, yet was derived from a single bone-marrow cell. "It proved for the first time that there was such a thing as a stem cell," said McCulloch.

Diseased bone marrow such as this can be treated with injections of healthy marrow cells.

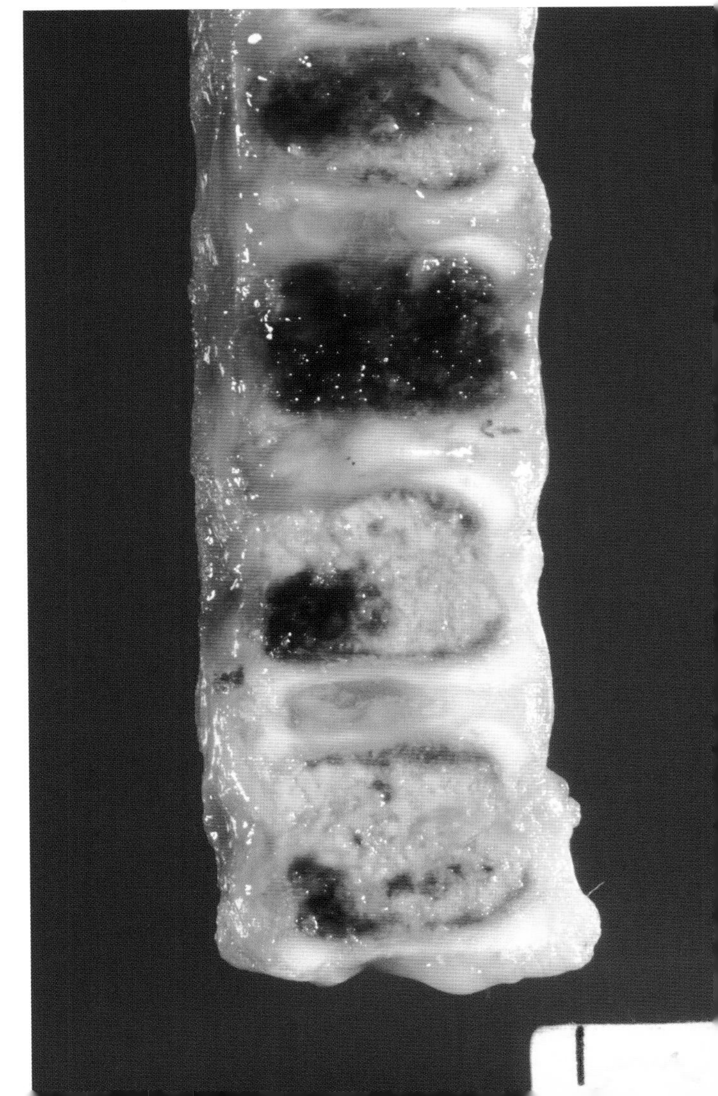

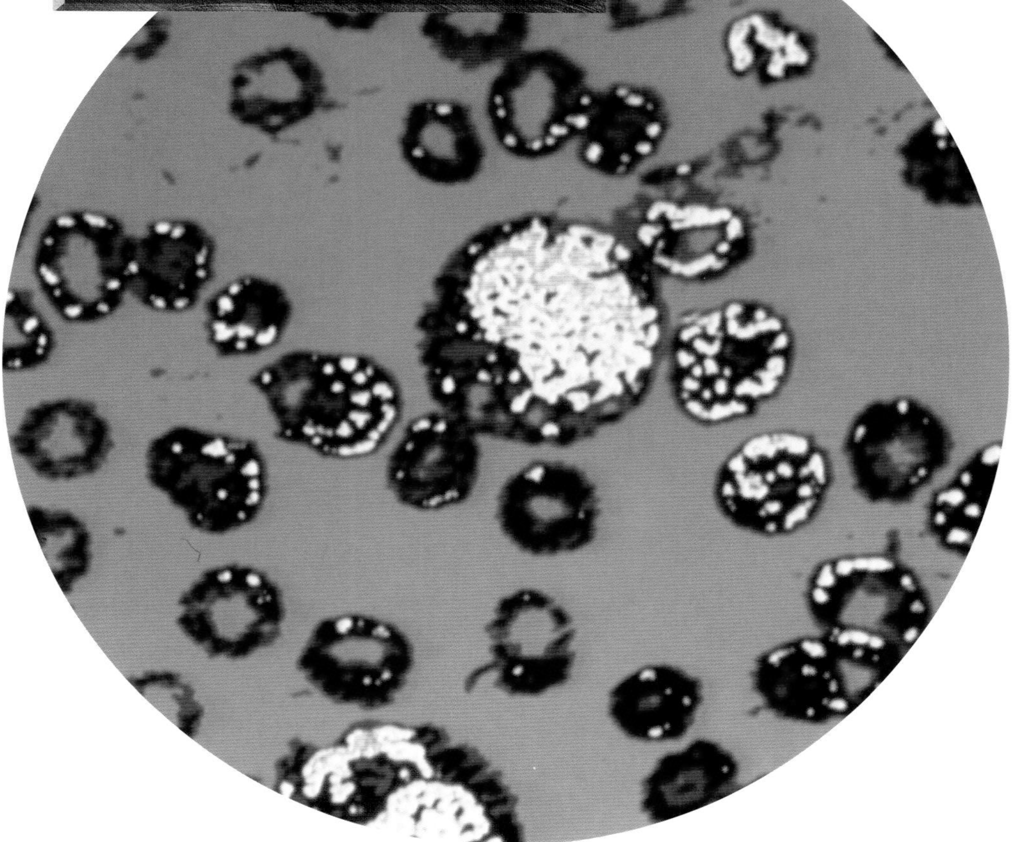

Blood cells affected by leukemia (shown here) can be treated with an injection of hematopoietic stem cells.

These bone marrow cells are called hematopoietic (blood-forming) stem cells. After Till and McCulloch's 1961 report of the cells' existence, other researchers began to study the cells' potential medical uses. In 1968, doctors at the University of Minnesota performed the first bone marrow transplant in a human being. They infused whole bone marrow into the patient so that the hematopoietic stem cells it contained could replenish the person's blood supply. Today, hematopoietic stem cells are often used to treat people with blood diseases such as leukemia and anemia.

It was not until 1988, however, that hematopoietic stem cells actually were isolated from the pool of other bone marrow cells. Irving Weissman and his colleagues at Stanford University in California developed a way to separate stem cells from non–stem cells. To prove the few cells they isolated really were stem cells, they injected the cells into mice whose blood systems had been destroyed by radiation. The researchers found that as few as 30 of the cells could renew a mouse's entire blood supply.

Next, Weissman and his coworkers set out to find hematopoietic stem cells in humans. They could not expose humans to the same radiation experiments as mice, so they manipulated genes to create mice with a human immune system. By 1992, they reported that they had isolated what appeared to be human blood-forming stem cells. Beginning in 1996, these cells were tested in patients with various kinds of cancer.

Stem cells taken from the spinal cords of rats have played a pivotal role in research.

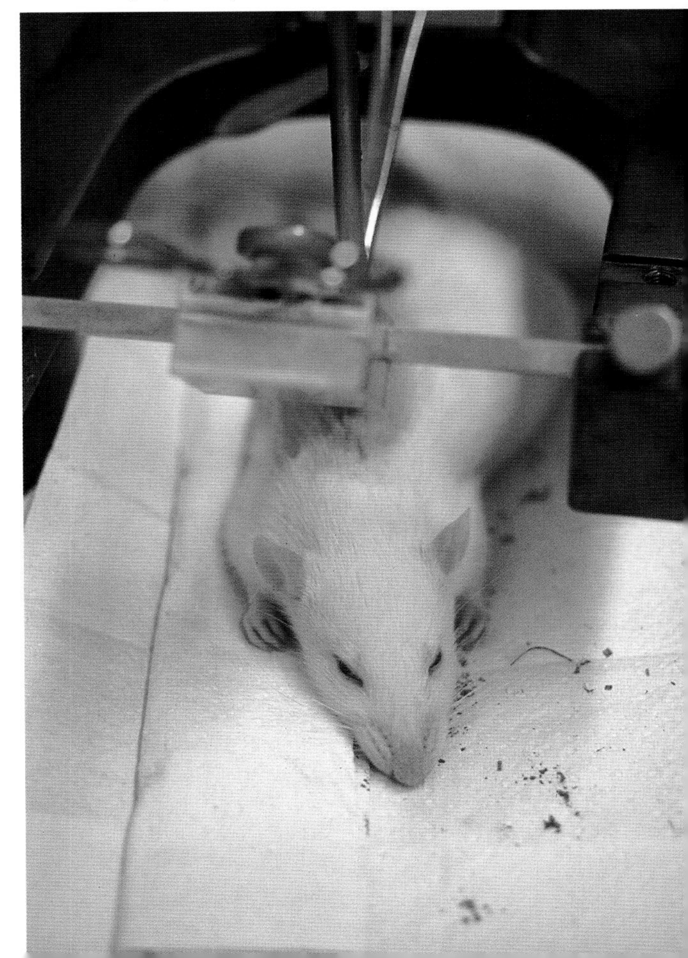

One test involved women with late-stage breast cancer that had spread through their bodies. This type of cancer is treated with chemotherapy, which kills cancer cells. Chemotherapy also destroys healthy cells, though, including hematopoietic stem cells in the bone marrow. The scientists decided to find a way around this problem. Before the women received chemotherapy, the researchers removed some of their bone marrow and purified the stem cells. After chemotherapy was complete, the scientists put the purified stem cells back into the women's bodies. More than four years later, about half the women remained free of cancer. This was a remarkable improvement over earlier treatments for late-stage cancer, which is extremely difficult to treat.

This microscopic view shows breast cells affected by cancer.

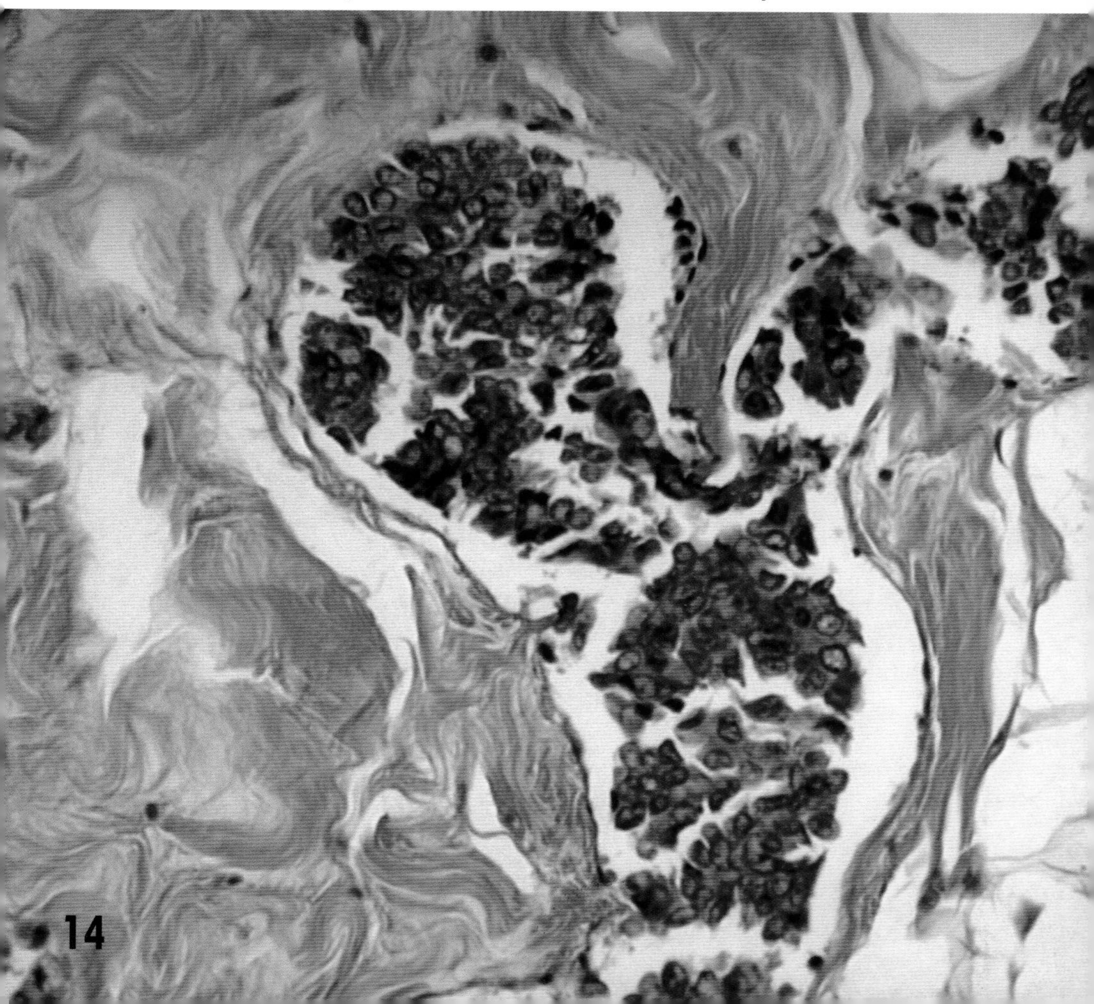

STEM CELLS EVERYWHERE?

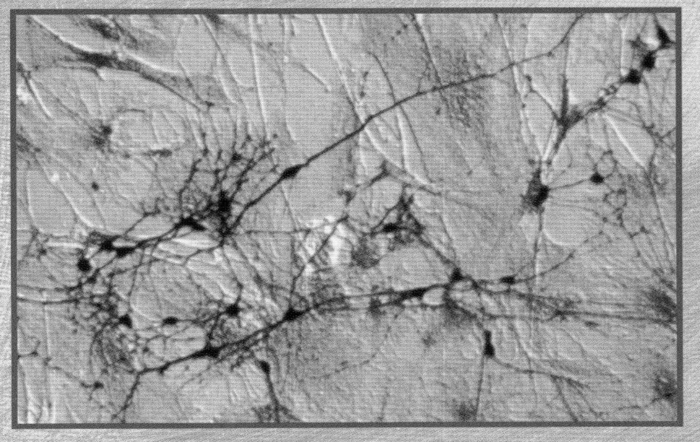

Nerve connections in the brain cells of a mouse embryo

At one time, scientists believed that only certain parts of an adult body contained stem cells. They believed that stem cells existed only to replace cells with short life spans. For example, cells that line the inside of a person's stomach generally live about three days before they are killed by acid in the stomach. These cells must be replaced continually; otherwise the person becomes ill and dies.

Stem cells have not yet been found in all parts of the body, but they have been identified in many more tissues than scientists once thought possible. Most surprising was the discovery of stem cells in the brain. It was long thought that the adult brain and spinal cord could not generate new cells—that the neural stem cells disappeared before birth, after the brain and spinal cord had formed.

In 1992, however, researchers at the University of Calgary in Canada found cells in the brains of mice that can divide to produce new cells. They isolated the cells and grew them in a laboratory dish. In 2002, researchers reported that stem cells isolated from the brains of adult rats can mature into working nerve cells. This suggests that at some time in the future it may be possible to use such stem cells to replace damaged or destroyed brain cells.

WORKING WITH EMBRYOS

At the same time that some scientists were proving the existence of stem cells in adults, other scientists were trying to find stem cells in very young embryos. The first discoveries of stem cells in embryos were reported in 1981 by two groups of researchers working with mice. One group, led by Gail Martin, was at the University of California in San Francisco. The second group, led by Martin Evans, was at the University of Cambridge in England.

Gail Martin explained that she removed tiny embryos just 3 1/2 days old from mice. She used special chemicals to separate the suspected stem cells. She placed the cells in laboratory dishes that contained substances essential for growth. Then she placed the dishes in incubators set at 99°F (37°C). A week later, she saw tiny colonies of cells growing in the dishes. "There have been a couple moments in my career when I've been really excited, and one was the night I saw those stem cells growing," Martin recalled.

It was Martin who gave the name "embryonic stem cells" to these cells. The name is now used by researchers around the world, and sometimes abbreviated as ES cells. In contrast, blood-forming stem cells and other stem cells found in the body after birth are called adult stem cells, sometimes abbreviated AS cells.

This mouse embryo has just divided for the first time from a single cell to a double cell.

The discovery of mouse embryonic stem cells had a big impact on research. It increased scientists' ability to study how embryos develop. It also helped to revolutionize the way that scientists study the functions of genes—the units of heredity that determine the characteristics of an organism. Scientists can insert a specific gene into a mouse embryonic stem cell, grow the stem cell into an embryo, and study the effect of the gene.

This animal was the first genetically modified rhesus monkey ever born.

Scientists began to look for—and find—stem cells in other kinds of animals. They steered clear of looking for human embryonic stem cells, though. They were concerned about the ethics of destroying human embryos for research.

By the mid-1990s, however, the potential value of understanding human embryonic stem cells was increasingly obvious. An important milestone occurred in 1996, when researchers at Indiana University demonstrated that mouse embryonic stem cells could be chemically induced to become heart cells, then implanted into—and become a functioning part of—a mouse heart. Suddenly, scientists realized that the cells could be used not only to increase scientific knowledge but possibly to repair damaged hearts and for other medical purposes.

In 1995, James Thomson and colleagues at the University of Wisconsin derived embryonic stem cells from rhesus monkeys,

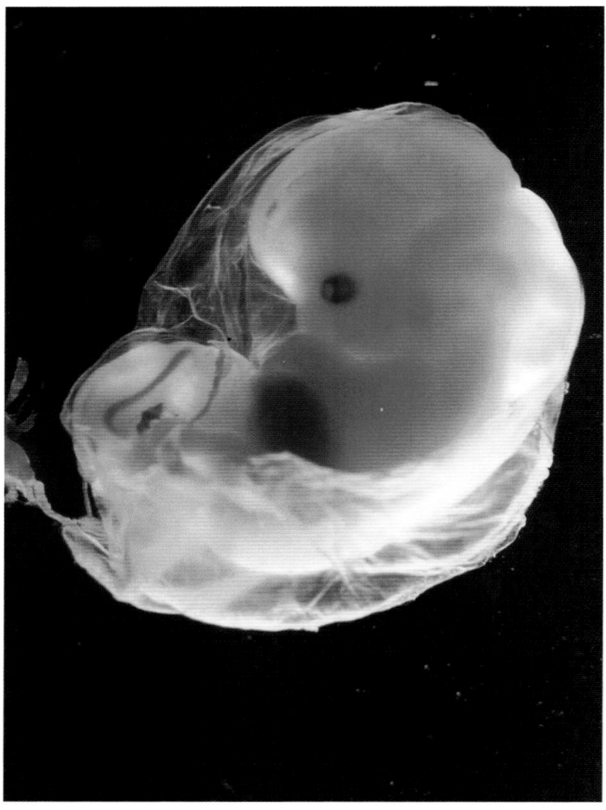

Human embryo

which are closely related to humans. Thomson decided to look at human embryos. "I just decided it would be important enough to do it," he later said.

Thomson obtained embryos created in a fertility clinic by couples that planned to use them to have children. Often, many more embryos are created than used. The extra embryos are considered to be surplus, and when the couples no longer want to use them, they usually are discarded. Some couples donated their surplus embryos to Thomson for his research. In 1998, he reported that his team had successfully isolated human embryonic stem cells and grown them in laboratory cultures.

Thomson's discovery was widely reported on television and in newspapers and magazines. Suddenly, "stem cells" became part of everyday vocabularies. With this came discussions of the potential uses of human embryonic stem cells, and debates about the pros and cons of such use. The discussions and debates continue today, against a backdrop of frequent reports of new ways in which scientists are learning about stem cells of all kinds.

RESEARCH MARCHES ON

As the 20th century gave way to the 21st century, growing numbers of researchers in the United States and other countries were manipulating stem cells in ways that could possibly lead to valuable medical applications. Scientists at Osiris Therapeutics, a company in Baltimore, Maryland, reported that they were able to coax certain bone-marrow stem cells to form not only blood cells but also bone, cartilage, and fat cells. A group led by Marc Hedrick at the University of California at Los Angeles found that human fat appeared to be an excellent source of stem cells. Thomson and his colleagues reported that they had changed human embryonic stem cells into blood-making cells. Researchers in Israel reported that they had transformed human embryonic stem cells into cells that produced insulin, a hormone normally made by cells in the pancreas.

Human fat cells, shown here, have been determined to be an excellent source of stem cells.

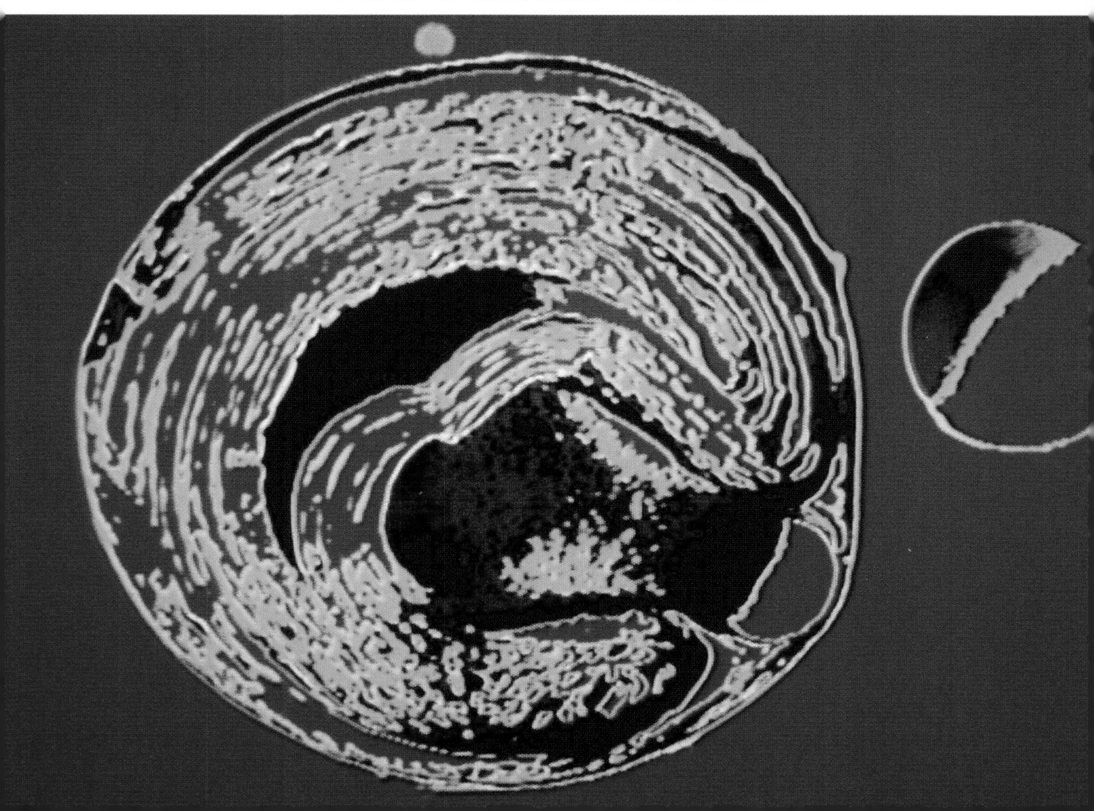

PLANTS HAVE STEM CELLS, TOO

Lilies, maple trees, carrots, and rose bushes look very different from one another. Like all plants, however, they begin life as embryos. Unlike animals, almost all their development takes place after the embryo stage. A plant body consists of roots, stems, and leaves. These parts are derived from stem cells—called "initials" by some plant scientists. New roots develop from stem cells at the tip of existing roots, in a region called the root apical meristem. New aboveground organs develop from stem cells at the tip of shoots (young stems), in a region called the shoot apical meristem. The next time you look at a plant, notice that the smallest leaves are at the ends of the stem. These are the youngest leaves; the largest—and oldest—leaves are at the bottom of the stem.

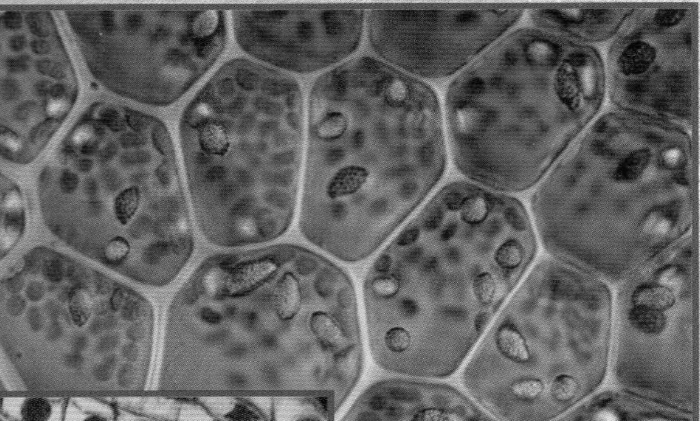

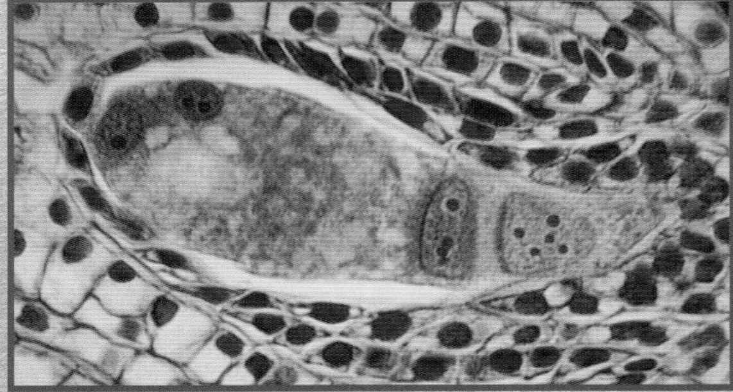

Plant cells

The stem cells continuously divide. They give rise to two kinds of cells: new stem cells—which enable a plant to grow throughout its life—and cells that go on to differentiate into the specialized cells of roots, stems, or leaves.

CHAPTER 2

STEM CELL RESEARCH TODAY

To build on the progress made in stem cell research so far, scientists are looking at the very earliest stages of an organism's development. They want to understand exactly how stem cells form and how they can be identified. The scientists also want to be able to grow stem cells in large quantities in the laboratory, and learn how to direct stem cells along specific paths of differentiation.

DEVELOPMENT OF AN ORGANISM

In humans and most animals, fertilization is the first stage in the development of a new individual. Fertilization is the union of an egg cell produced by the mother and a sperm cell produced by the father. The result is a single cell called the fertilized egg.

Within 24 hours of fertilization, the fertilized egg divides into two identical

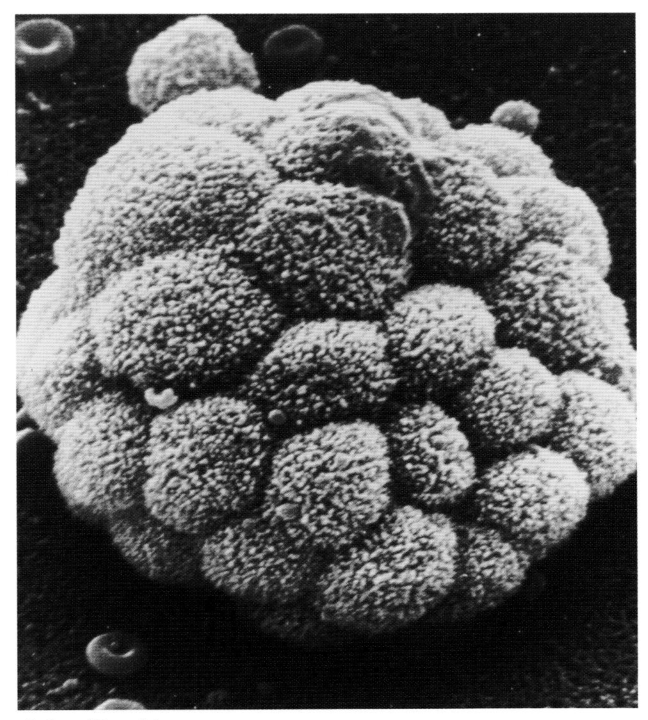

A fertilized human egg

This light micrograph shows a four-cell human embryo two days after fertilization.

cells. This is the beginning of the embryo. A series of rapid divisions follows: the 2 cells become 4 cells, then 8 cells, 16 cells, 32 cells, 64 cells, and so on. The cells form a hollow ball called a blastocyst. Inside, at one end of the ball, a cluster of cells develops. This cluster is called the inner cell mass.

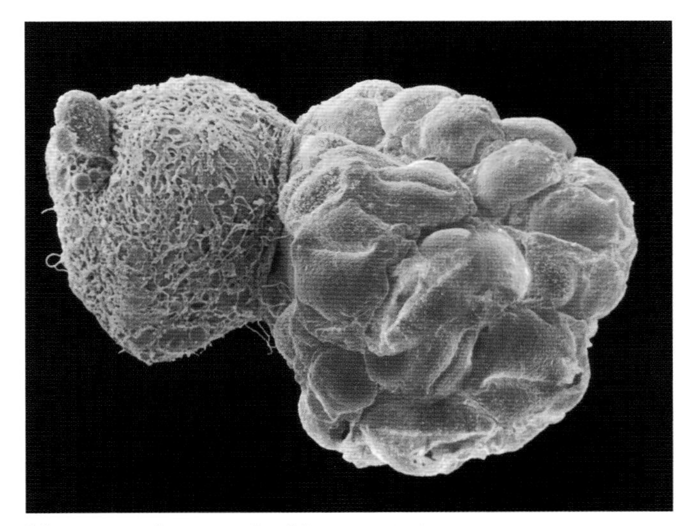

Human embryo at the blastocyst stage, five days after fertilization

At this point, a human embryo is 4 to 5 days old and consists of 200 to 250 cells. If it becomes implanted in the mother's uterus, the outer cells of the blastocyst will form the placenta—a membrane that carries substances between the mother and the developing young. The cells of the inner cell mass will, over a period of about 9 months, form the infant. Approximately 30 in number, these are the embryonic stem cells.

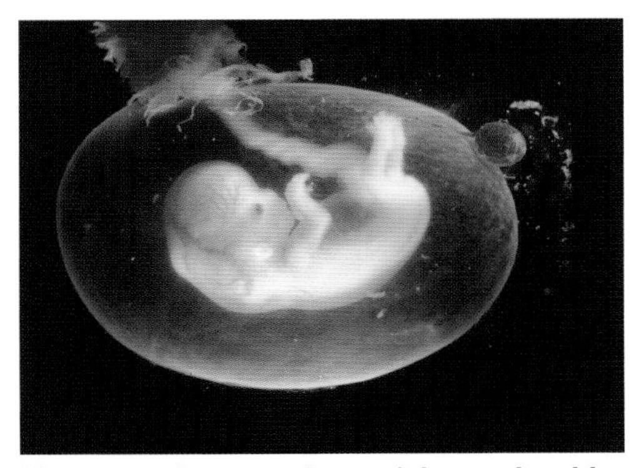

Human embryo at about eight weeks old

When the embryo is 14 to 16 days old, the embryonic stem cells react to genetic instructions and begin to form three layers: the ectoderm, mesoderm, and endoderm. The cells in these layers still appear more or less similar in structure. The three layers, however, gradually differentiate. Over the following months, they give rise to all the different tissues and organs in the body. The ectoderm ("outer skin") forms the skin, hair, nervous system, and lining of the mouth. The mesoderm ("middle skin") forms the bones, most muscles, kidneys, heart, blood vessels, blood, and reproductive organs. The endoderm ("inner skin") forms the respiratory system, liver, pancreas, and lining of the digestive system.

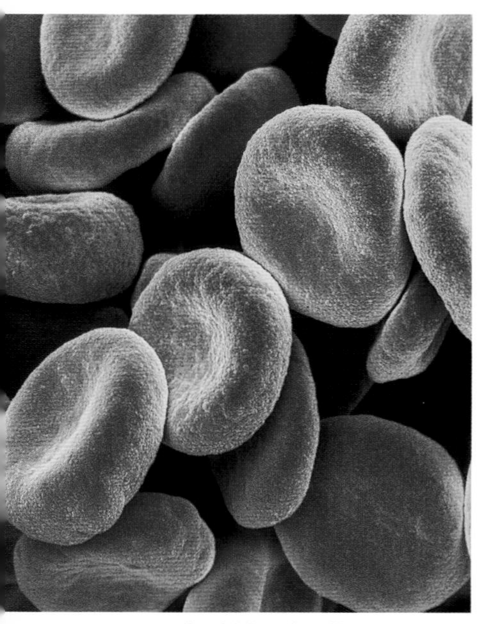
Red blood cells

At birth and throughout life, the body consists almost entirely of mature cells, which have a limited life span. Therefore, the body's existence depends on a very small number of stem cells in each tissue. More specialized than the embryonic stem cells, these are called adult stem cells.

Adult stem cells give rise to replacements for mature cells that have died or been damaged by injury or disease. For example, the outer cells of the skin are continually being rubbed off. Red blood cells, which carry oxygen throughout the body, last only about two months. White blood cells fight infections and constantly die in battles with germs. Even the bone cells of the skeleton are replaced about every seven years.

To learn the habits of stem cells and to understand how embryonic and adult stem cells differ from one another—and from specialized cells—scientists need to isolate the stem cells and grow them in the laboratory. The first step in the process is to identify the stem cells.

MORE CELLS, MORE CONTROVERSY

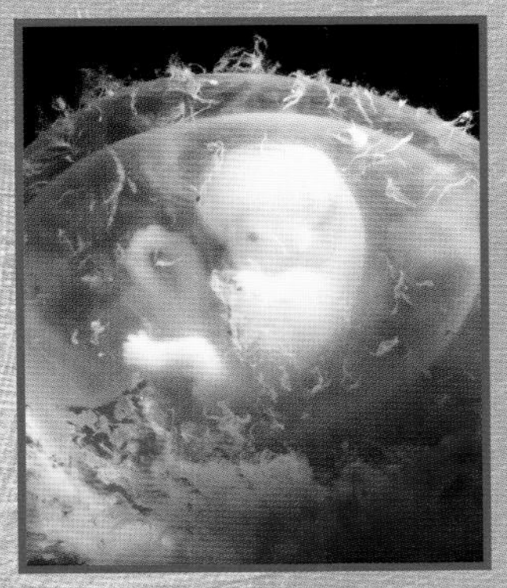
Human embryo

When a human embryo is about six days old, it attaches itself to the wall of its mother's uterus. Soon thereafter, a small group of the embryo's stem cells migrates toward the embryo's developing ovaries (egg-producing organs in females) or testes (sperm-producing organs in males). Known as embryonic germ cells, these cells are destined to be part of the reproductive system. Scientists who have studied these cells have found that they appear to have many of the same properties as embryonic stem cells.

Embryonic germ cells, however, begin to differentiate later than embryonic stem cells do. For that reason, they can be obtained only from aborted embryos. Because abortion is such as divisive issue, the use of embryonic germ cells in research is even more controversial than the use of embryonic stem cells.

Some people oppose research on both embryonic stem cells and embryonic germ cells because it involves the destruction of embryos. They believe that human embryos are human beings. Destruction of a human embryo—whether in a laboratory dish or during an abortion—is murder, they say. Other people who are opposed to abortion are willing to support research on embryonic stem cells but not on embryonic germ cells.

People who support the research feel differently. They do not consider it wrong to use embryos that would otherwise be discarded by fertility clinics or abortion clinics. Rather, they believe it would be immoral not to try to use the cells to help improve the quality of life for people who have been crippled in accidents or who are suffering from deadly illnesses.

IDENTIFYING STEM CELLS

The history of stem cell research demonstrates that it is not an easy task to identify stem cells. Unlike nerve cells, blood cells, and other mature cells, stem cells do not have a distinct appearance. Therefore, researchers define stem cells not by their shape or size but by their ability. They periodically test the cells they culture (grow) to determine if the cells exhibit the proper characteristics.

Embryonic stem cells removed from an embryo and grown in laboratory cultures have a very important characteristic: they can divide and renew themselves for long periods of time. Under proper conditions, the cells divide over and over again to form more embryonic stem cells. This process can continue for a year or even longer. At intervals, the scientists will examine the cell cultures through a microscope to see if the cells look healthy and if they still have not differentiated. One important clue is the protein Oct-4, which typically is made by embryonic stem cells. This protein helps turn genes on and off at the right time. It disappears after embryonic stem cells begin to differentiate.

Cells in a laboratory culture that have tentatively been identified as embryonic stem cells can be removed from the culture and grown in a new culture, or subculture. If the original cells were indeed embryonic stem cells, the subculture will consist of undifferentiated cells exactly like those in the original culture. A whole series of subcultures can be made, one after the other.

It is more difficult to identify, isolate, and grow adult stem cells in the laboratory. One problem is that adult stem cells exist in small numbers and are surrounded by other kinds of cells. They cannot easily be separated from the non–stem cells. Irving Weissman, whose Stanford University team were the first scientists to isolate adult stem cells, found that in the bone marrow of mice, there is one stem cell for every 10,000 bone marrow cells.

One method used to identify adult stem cells is to label sample cells with a chemical marker. The marker is inserted into the cells' genetic material, which allows it to be passed down from one generation to the next. If the cells then produce several kinds of cells, and these cells all contain the chemical marker, that means the original cells are stem cells. For example, if cells removed from bone marrow divide to produce red blood cells plus different kinds of white blood cells, and all the new cells contain the chemical marker, then the original cells were hematopoietic stem cells. If cells removed from the brain and grown in a culture give rise to nerve cells plus two other kinds of cells found in nerve tissue, then the original cells were neural stem cells. If, however, an original cell does not develop into a variety of specialized cells, it was not a stem cell.

Stem cells can be collected in a hospital or laboratory setting and then identified for later use.

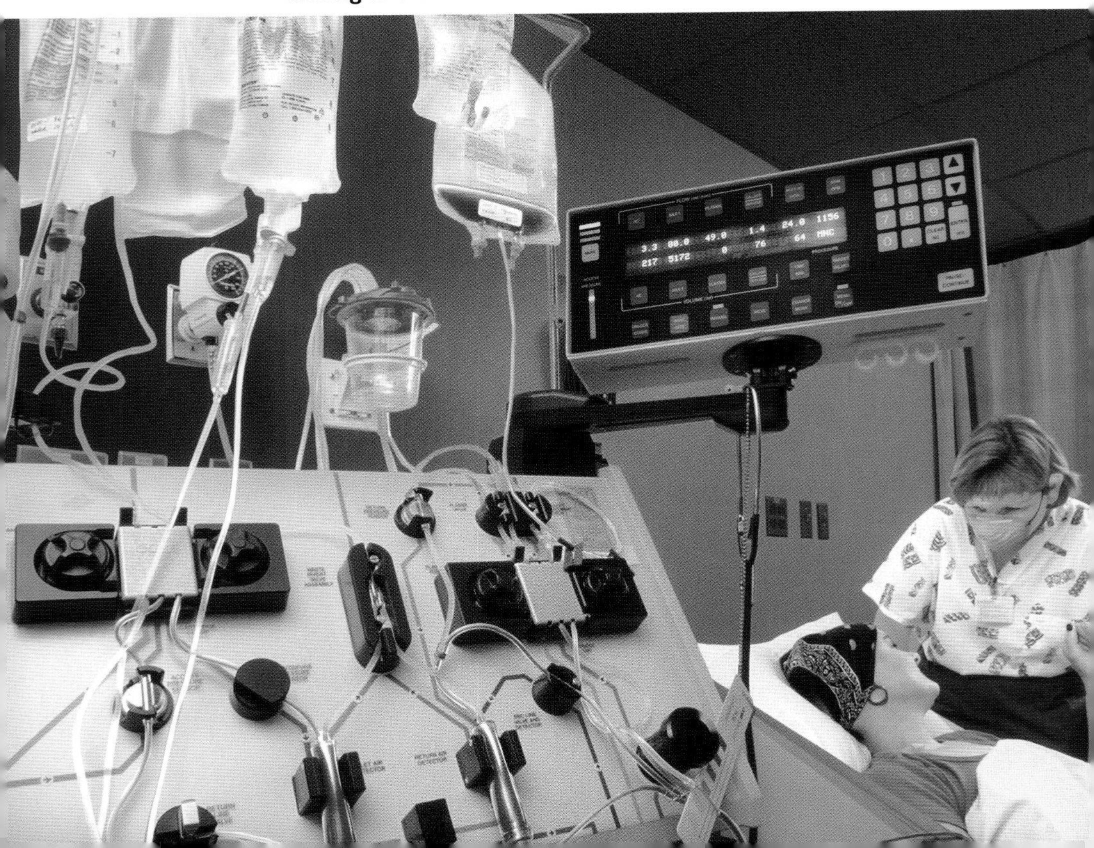

CELLS FROM THE UMBILICAL CORD

As a human embryo develops within its mother's uterus, it is connected to the placenta by an umbilical cord. The embryo's blood flows through blood vessels in the umbilical cord and in the placenta picks up nutrients and oxygen from the mother's blood. There is no direct connection between the blood supplies of the embryo and mother. Rather, the exchange of materials takes place through the walls of the blood vessels.

Human embryo attached to placenta

At the time of birth, the umbilical cord is cut and discarded. Blood extracted from the detached umbilical cord a few minutes after birth contains a unique kind of stem cell. Scientists have discovered that these stem cells can be used instead of a bone marrow transplant to help patients who have leukemia or certain other cancers. The umbilical cord stem cells speed up recovery of a patient's bone marrow and blood following chemotherapy.

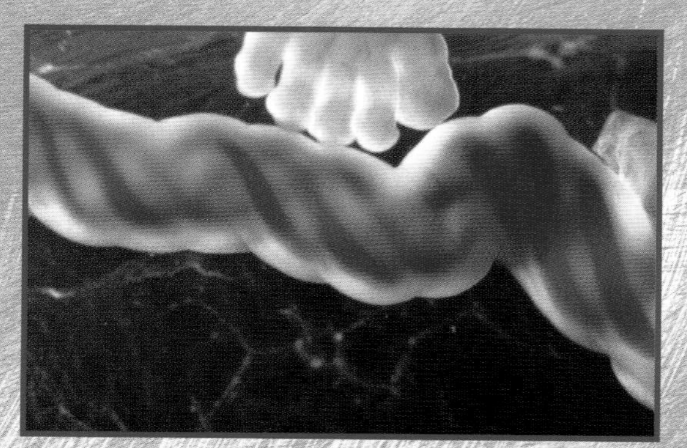

Close up of human umbilical cord at about ten weeks

GROWING STEM CELLS IN THE LABORATORY

Stem cells studied and grown in laboratories come from different sources. Adult stem cells come from living organisms. A tiny skin sample may be removed from a mouse in an attempt to culture mouse skin stem cells. A tiny sample of leg muscle may be removed from a human in hopes of growing human muscle stem cells.

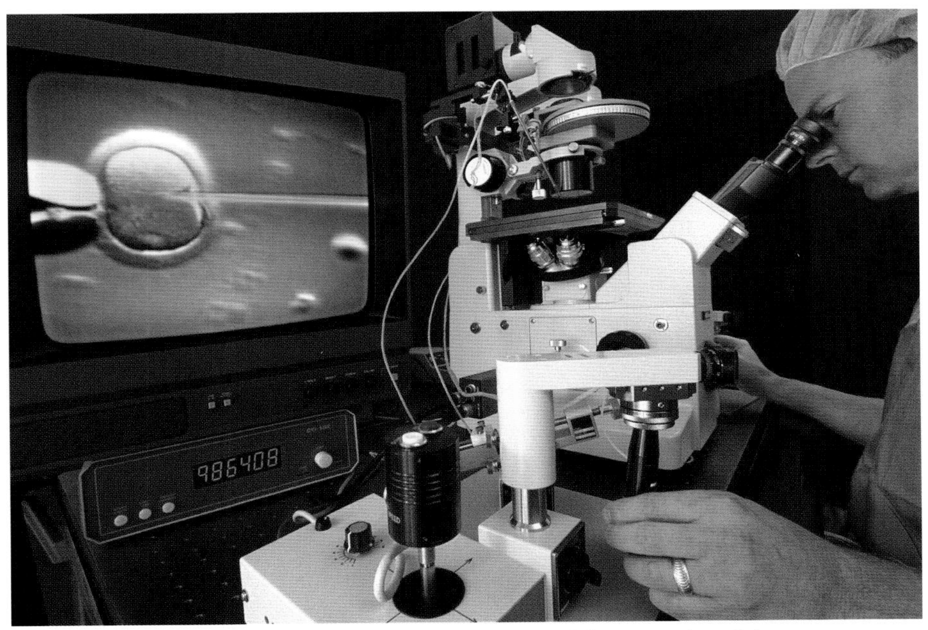

At fertility clinics, human eggs are united with sperm in a laboratory dish to create fertilization. Fertility clinics provide embryos that are used in stem cell research.

Human embryos used in stem cell research come mainly from fertility clinics. The purpose of these clinics is to try to help couples that have been unable to achieve pregnancies. At a clinic, the woman's eggs are united with the man's sperm in a laboratory dish. Ten to twenty eggs may be fertilized at a time. Each is grown into an 8-cell embryo. Then several are placed into the woman's uterus. If all goes as hoped, at least one of the young embryos will develop into a baby.

Spare embryos can be frozen for later use.

A couple is likely to use some but not all of the embryos in efforts to produce pregnancies. The spare embryos are frozen in liquid nitrogen. Until recently, a couple could choose to keep them in storage or ask that they be discarded. Now, in some countries, couples have a third choice: With their agreement, the embryos can be given to scientists for research.

A scientist's first step is to gently thaw the 8-cell embryos. Then the embryos are grown in the laboratory until they have developed into blastocysts. The next step is to carefully remove some cells from each blastocyst's inner cell mass and try to turn them into a line of self-reproducing stem cells. This is known as cell culture.

Cell culture typically is done in a petri dish, a round, shallow plastic or glass dish with a lid. The dish is sterilized to ensure that it is free of germs and other contaminants. Then a mixture of chemicals is poured into the dish. This mixture is called the medium. It contains amino acids (the building blocks of proteins) and other food

Laboratory petri dishes are filled with various cell cultures.

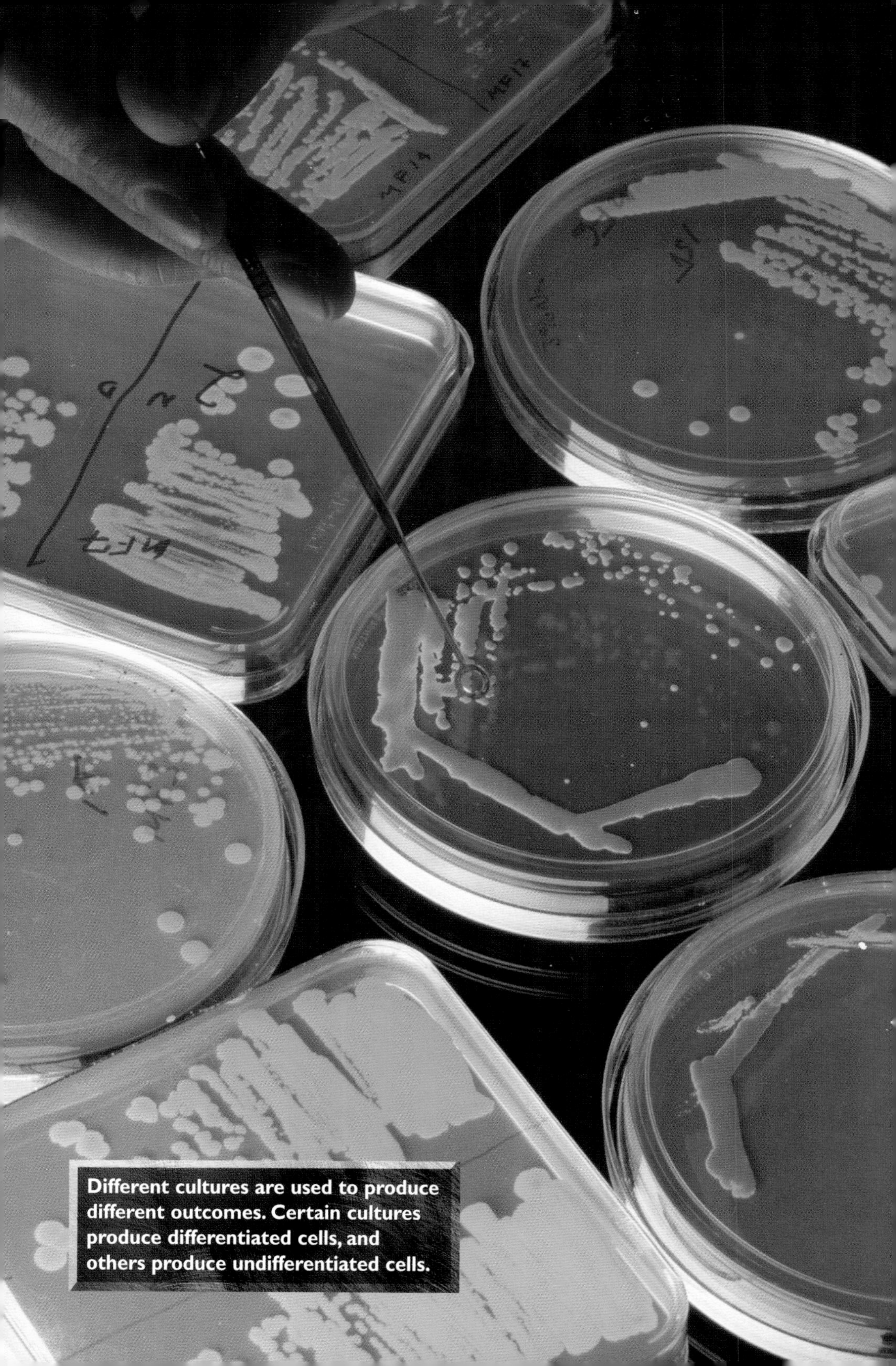

Different cultures are used to produce different outcomes. Certain cultures produce differentiated cells, and others produce undifferentiated cells.

and growth substances. It may also contain antibiotics, such as penicillin and streptomycin, to protect against bacteria.

The composition of culture mediums varies depending on the researchers' objectives. For instance, what works for growing undifferentiated embryonic stem cells differs from what is needed to cause the cells to differentiate. Making the cells develop into muscle cells requires a different mix of chemicals than does making them develop into skin cells. The concentrations of the different chemicals may affect the growth rate of the stem cells; that is, the cells may reproduce faster in one mixture than in another that contains the same chemicals but in different proportions.

In addition, the culture medium used to grow stem cells of one type of organism may not work for other kinds. For example, the culture medium traditionally used to grow mouse embryonic stem cells cannot be used to grow rabbit or human embryonic stem cells.

In many cases, scientists have not yet determined the proper mix of nutrients needed to achieve certain results. One of their major goals is to perfect techniques that will allow them to coax all the cells in a colony of embryonic stem cells to develop into a specific kind of adult cell, such as hair cells or stomach lining cells or insulin-producing cells.

Another important goal is to ensure that colonies of stem cells proliferate indefinitely. Once a colony of stem cells has been established in a dish, individual cells can be transferred to other dishes. Under proper conditions, each cell will divide to produce two cells that are exact copies of the original stem cell. Repeated divisions result in the formation of a subcolony that spreads over the surface of the dish. By repeating this process again and again, it is possible to obtain millions of genetically identical embryonic stem cells from the original cells of a blastocyst's inner cell mass.

CHAPTER 3

THE PROMISE OF STEM CELLS

In recent years, numerous scientists in laboratories throughout the world have made stem cells the focus of their research. There are three main reasons for this interest in stem cells: their value for basic research, the prospect of using them to treat illness and injury, and the possibility of using them to develop new medicines.

Many scientists believe that stem cells hold the answers to important new medicines and treatments that can reduce human suffering.

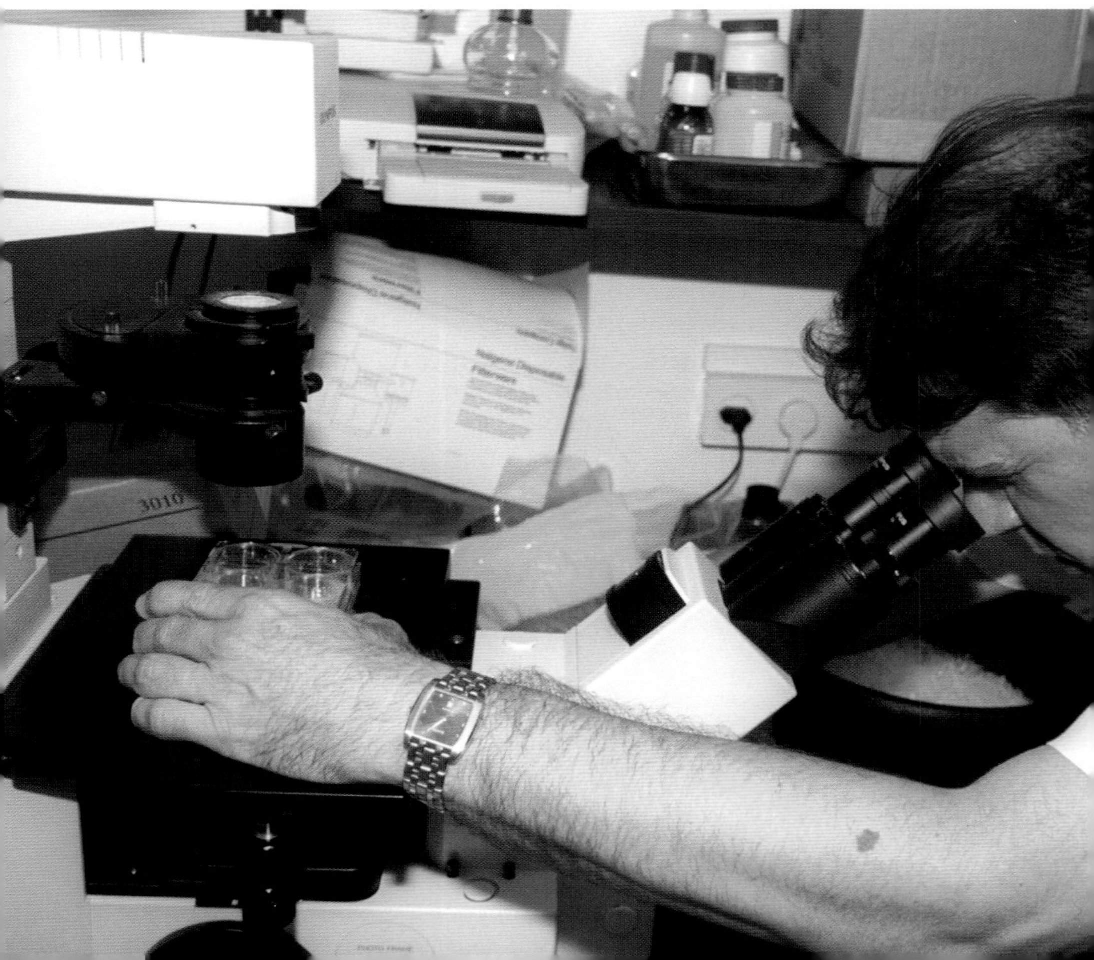

UNDERSTANDING LIFE

As in other sciences, new discoveries about stem cells lead to new questions. At one time, scientists asked whether an embryo contains special cells that form all the different cells in the adult organism. After they discovered embryonic stem cells, new questions arose: What genes play a role in giving stem cells their special ability? How do the cells change into specialized cells? What genes or other factors direct these changes? Why do some stem cells develop into arms and others into livers? How does the process that produces a nerve cell differ from the process that results in the formation of a blood cell?

One scientist who is tackling such questions is Douglas Melton at Harvard University. He began to study stem cells after his son was diagnosed with type I diabetes. Diabetes is a condition in which a person's body does not produce enough insulin, a hormone that is made in the pancreas and is essential to breaking down carbohydrates. Melton wants to understand how stem cells make a pancreas. He explains, "We want to know what genes and cells are involved in each decision so we can learn how to direct the cells' differentiation down that pathway."

Deciphering all the genetic and cellular events that occur during human development will help Melton and other scientists understand what goes wrong to cause diseases and conditions such as birth defects. Some of the most serious medical conditions result from abnormal cell division and differentiation. Cancer is an example. Normal cells have genes that tell them when to stop dividing. Cancer cells, by contrast, divide uncontrollably. If scientists can understand what turns genes on and off in stem cells, they may be able to use this knowledge to understand what causes normal cells to become cancer cells.

Scientists also want to know how much difference there is among embryonic stem cells of different people. Is there a lot of variety or not very much? The way to answer this question is to study stem cells from people who represent a broad cross-section of the world's

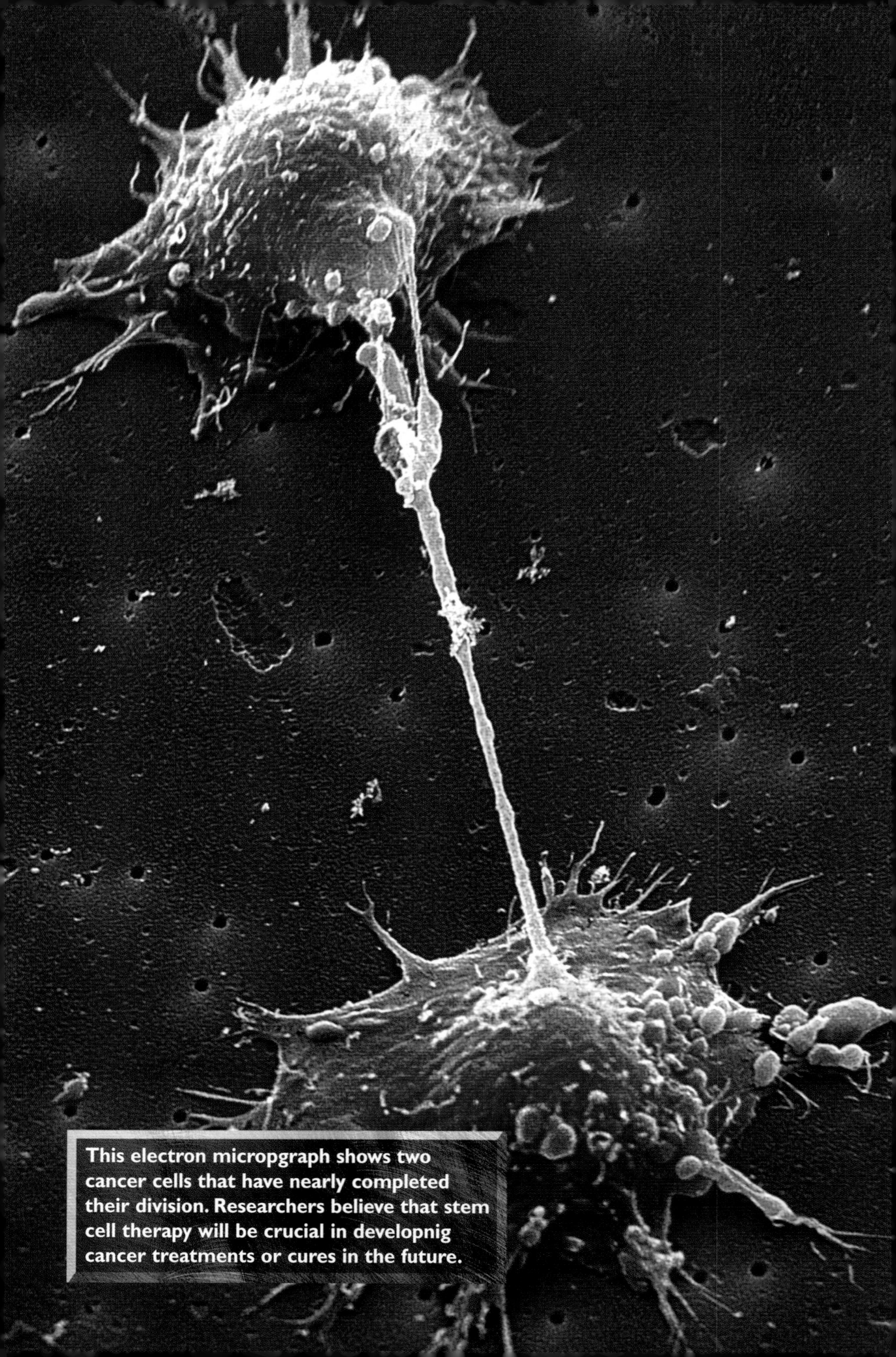

This electron micropgraph shows two cancer cells that have nearly completed their division. Researchers believe that stem cell therapy will be crucial in developnig cancer treatments or cures in the future.

population. Diversity also is important to ensure that stem cells are obtained from people who have inherited tendencies for various diseases. Researchers want to learn if the embryonic stem cells of people who inherit a high risk of a disease differ from those of people at low risk.

There also are many questions to be answered about adult stem cells. These include: How many kinds exist and where are they located? Do they develop directly from embryonic stem cells or in some other manner? What signals regulate their differentiation? Does an adult stem cell have the potential to act as an embryonic stem cell and give rise to all kinds of specialized cells?

TREATING AND CURING ILLNESS AND INJURY

Both embryonic and adult stem cells hold great promise for treating diseases and injuries. Among the diseases believed to be treatable with stem cells are heart disease, type I diabetes, cancer, and Parkinson's disease. Tissues damaged by burns or spinal cord injuries may also be reparable using stem cells.

This magnetic resonance image (MRI) shows a brain affected by Parkinson's disease. The arrows points to an area where nerve cells have been damaged or have died.

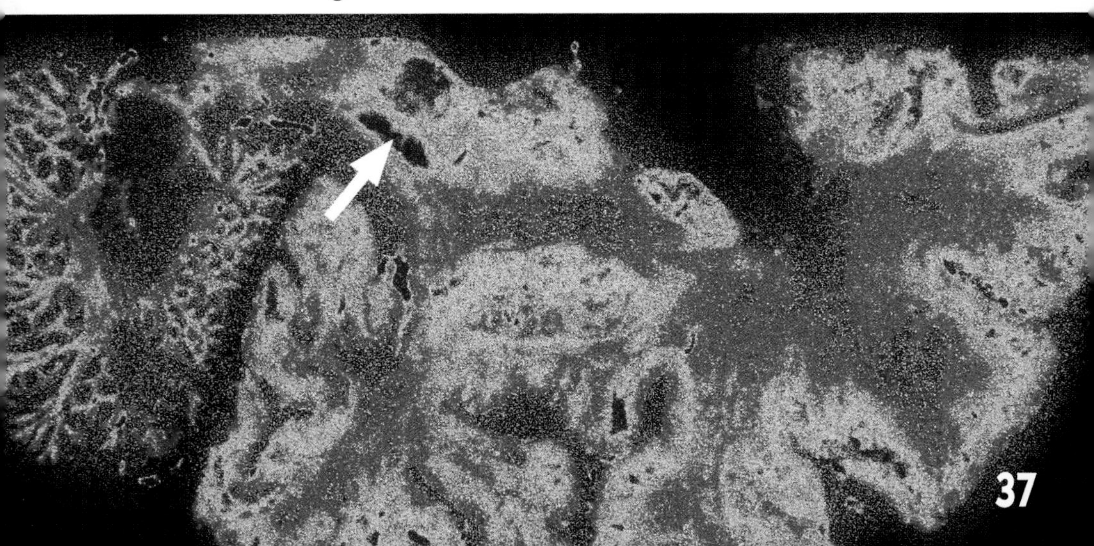

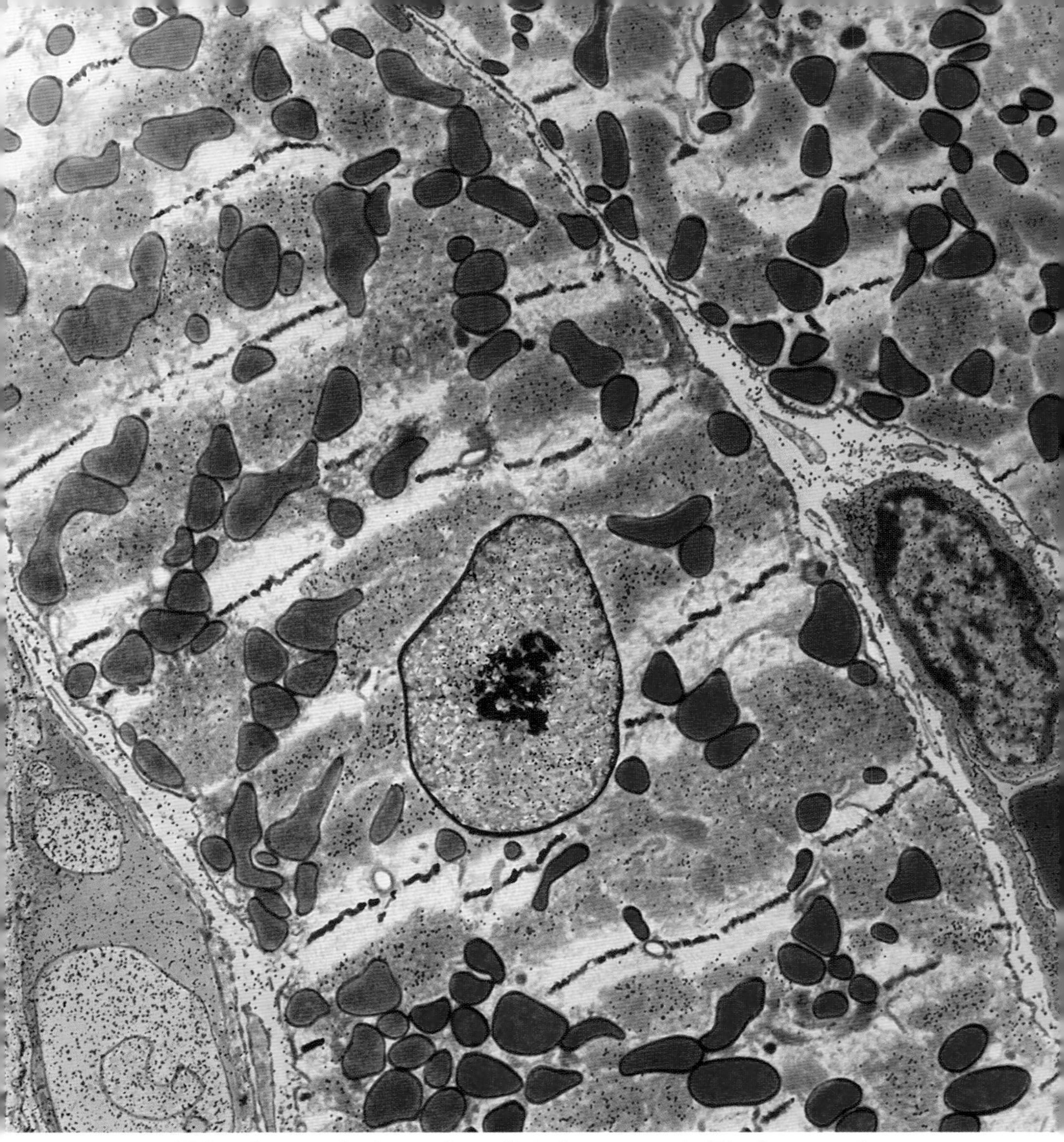

This micrograph shows the cells in heart muscle. The large nucleus is seen center.

To help a patient who has lost heart muscle due to heart attacks or other disease, researchers want to grow embryonic stem cells into heart muscle cells. These cells would then be injected into the patient's heart. They would help rebuild the heart muscle and restore good pumping action so that blood would move effectively through the body.

Embryonic stem cells grown in a culture medium that causes them to develop into pancreatic cells that produce insulin would be used to treat people with type I diabetes. When these cells are implanted into a patient, they would begin to produce insulin. This would cure the patient of the disease and eliminate the need for daily insulin injections.

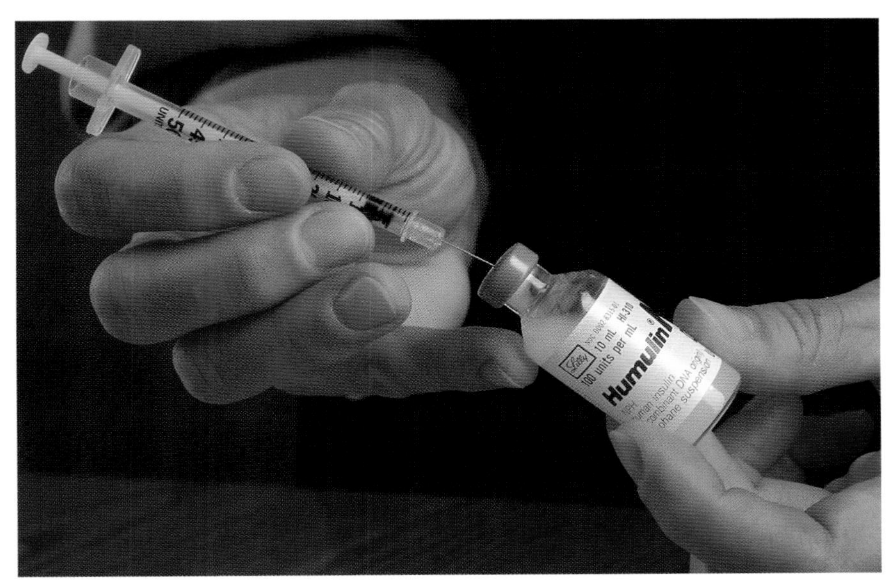

Most diabetics need to have regular injections of insulin. Stem cells can be grown into pancreatic cells that produce insulin, which could be used to treat and cure people with diabetes.

Patients with Parkinson's disease also might be cured thanks to embryonic stem cells. In this disease, nerve cells in the brain stop producing a chemical called dopamine. As a result, the nervous system loses its ability to control muscles. The patients experience muscle tremors, such as shaking of the fingers of one hand. Embryonic stem cells might be grown into nerve cells that produce dopamine. These cells would be transplanted into the patients and potentially reduce or eliminate tremors and other symptoms of the disease.

STEM CELLS AND CLONES

Dolly the sheep

A clone is an exact copy of a parent cell or organism. It contains the same genes as its parent. Strawberry plants produced by means of horizontal stems from a parent plant are clones. Some kinds of snails are among animals that naturally create clones of themselves.

Scientists have learned how to create certain clones in the laboratory. The best-known example was a clone named Dolly, a lamb born in 1996 in Scotland. She was created from a cell taken from the udder of a 6-year-old female sheep. Her genetic makeup was exactly the same as her mother's.

Dolly's birth caused people to fear that human clones could be next. Many countries passed laws that banned human cloning. Scientists point out, however, that there are two kinds of cloning. Reproductive cloning leads to the creation of a new organism. Therapeutic cloning produces cells that can be used to study and perhaps treat disease.

In the cloning process called nuclear transfer, the nucleus is removed from an unfertilized egg. The nucleus from an adult cell is injected into the egg. The egg is grown into a blastocyst, from which researchers can remove embryonic stem cells and prompt them to grow into specialized cells.

Irving Weissman at Stanford University described how this process might be used to study Lou Gehrig's disease, which is caused by the death of motor neurons (nerve cells that control the muscles). A person with Lou Gehrig's disease has one or more faulty genes that result in the disease. The nucleus from one of the person's cells—which contains his or her genes—could be transferred into an egg cell. The resulting embryonic stem cells could be exposed to chemical signals that would cause them to develop into motor neurons exactly like those in the person's body. Scientists could study these cloned neurons to locate the faulty genes. Then drugs could be developed to counteract the genes' actions.

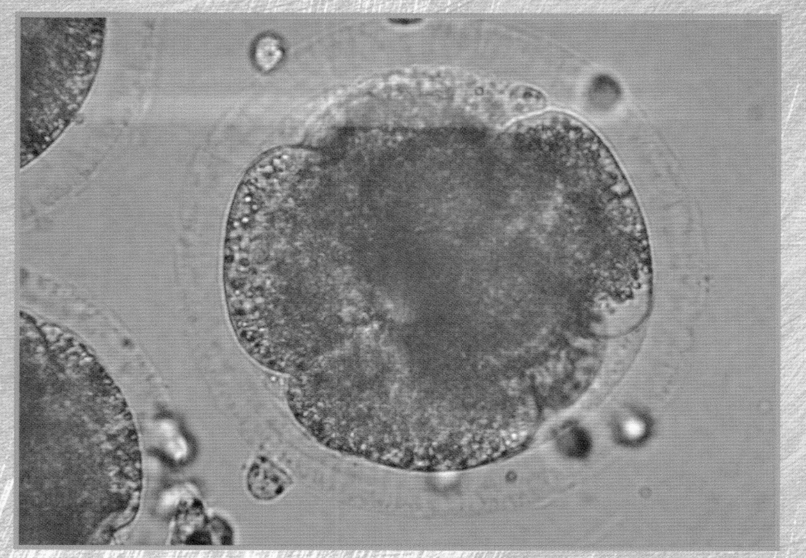

Cloned human embryo

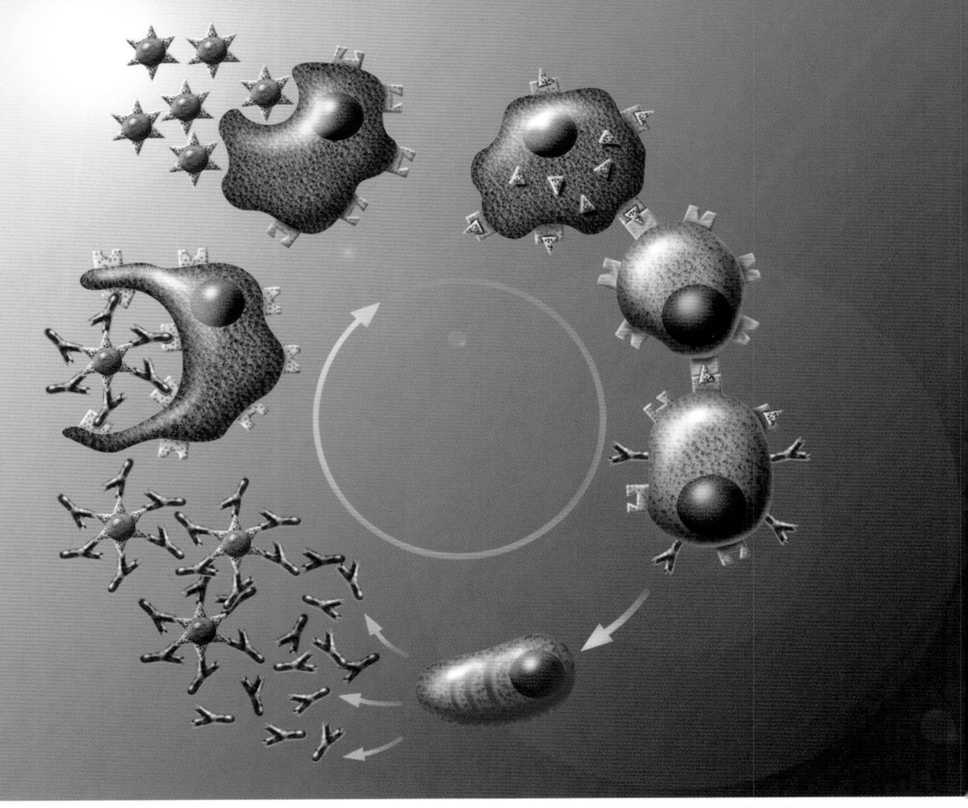

This illustration shows how the immune system works by defending against invaders. When specialized cells (blue) come into contact with bacteria (orange balls), they produce antibodies (y-shapes) that bind themselves to the bacteria. Once bound, the star-shaped bacteria are much more susceptible to capture by other cells that destroy the invaders.

Adult blood-forming stem cells from bone marrow already are being transplanted into patients to treat leukemia and other cancers. The cells are injected into the bloodstream and—in ways not yet understood—find their way to the patients' bone marrow, where they begin to differentiate into red blood cells and white blood cells. It is possible that other kinds of adult stem cells also could be used for transplants. For example, neural stem cells might replace dead or damaged nerve cells after spinal cord injuries.

To repair tissues, it may not even be necessary to use adult stem cells that normally give rise to those tissues. In 2001, researchers reported that they were able to use bone marrow stem cells to repair the heart muscle of mice that had had heart attacks. The researchers injected the stem cells into the hearts. Nine days later the stem cells were producing muscle cells.

One problem with human stem cell transplants that needs to be overcome is the fact that the human body does not like invaders. The body's immune system quickly attacks foreign cells—not only germs but also cells from other humans. This is a major obstacle to the transplantation of stem cells, tissues, and organs. A patient who receives a transplant usually must be kept on special drugs that suppress the immune system so it cannot reject the transplant.

A solution that may someday be possible would be to isolate and grow a patient's own stem cells. The stem cells could be taken from healthy tissue and then reprogrammed in the laboratory to develop into the desired specialized cells. These would then be transplanted into the patient. Since they developed from the patient's own cells, they would not be rejected.

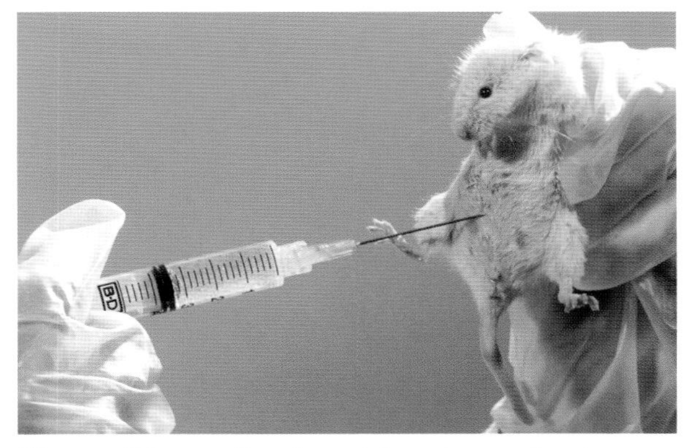

Stem cell harvesting from laboratory animals is important for effective research.

DEVELOPING NEW MEDICINES

Another area that may one day benefit from stem cell research is the testing of newly developed medicines. At present, experimental medicines are tested mainly in laboratory animals and, if they appear to be safe and beneficial, in human volunteers. This is a long, expensive, and sometimes dangerous process. About 90 percent of experimental medicines that are successful in animal tests fail when tested on humans, either because they are ineffective or because they have harmful side effects.

In the future, stem cells may change the way that drugs are developed and tested. Medicines could first be tested on skin, nerve, and a variety of other human adult stem cells to see if they harmed the cells or prevented them from producing specialized cells. Only if a medicine is shown to be safe and effective on stem cells, would it then be tested in animals and, later, in humans. This process would probably reduce the cost and length of time needed to bring new medicines onto the market.

In the future, stem cells may lead scientists and researchers to the creation of entirely new treatments and medicines. Because they can be used for testing, stem cells will also provide valuable information that will speed the process of discovery.

MUCH WORK STILL TO BE DONE

Scientists who study stem cells can follow many different research paths, all of which offer the possibility of exciting discoveries and applications. There are challenges to be met and problems to be solved along all of these paths. Scientists are optimistic, though, that at some point in the future humans will enjoy longer, better lives because of these uniquely versatile cells.

GLOSSARY

Adult stem cells Stem cells found in small numbers in the tissues of an organism after birth. Throughout life, they generate new specialized cells.

Cell The basic building block of an organism. A human is made up of trillions of cells.

Cell culture Growing cells in a laboratory.

Clone An exact copy of a parent cell or organism.

Differentiation The process of developing from unspecialized cells to specialized cells.

Embryo The early developing stage of an organism. In humans, the embryo is called a fetus beginning with its ninth week.

Embryonic stem cells Unspecialized cells found in an embryo from which all the specialized cells in the body develop. Often abbreviated as ES cells.

Gene A unit of heredity, found in the nucleus of a cell. It directs the production of an enzyme or other protein.

Nucleus The structure in a cell that contains the genes.

Placenta A structure in a pregnant woman's uterus that nourishes the developing embryo.

Stem cell An unspecialized cell that can generate specialized cells.

Therapeutic The use of medicines, stem cells, or other methods to treat diseases and disorders.

Tissue A group of cells of one or more types that work together to carry out a specific function.

Uterus The organ in a woman's body in which an embryo grows until the time of birth. Also called a womb.

FOR FURTHER INFORMATION

Books
Nardo, Don, *Cloning*. San Diego, CA: Blackbirch Press, 2003.

Viegas, Jennifer, *Stem Cell Research*. New York: Rosen Publishing Group, 2003.

Articles
Constans, Aileen, "Stem Cell Know-how: An Overview of Tools for Stem Cell Culture," *The Scientist*, September 2, 2002.

"Frankenstein Revisited," *Kids Discover*, July 2002.

Kolata, Gina, "The Promise of Therapeutic Cloning," *New York Times*, January 5, 2003.

Lewis, Ricki, "John Gearhart: Stem Cell Guru," *The Scientist*, December 9, 2002.

Wade, Nicholas, "In Tiny Cells, Glimpses of Body's Master Plan," *New York Times*, December 18, 2001.

Wade, Nicholas, "Study Expands Range of Stem Cell Abilities," *New York Times*, March 7, 2002.

Westphal, Sylvia Pagan, "Stem Cell Work Forges Ahead as the Politicians Squabble," *New Scientist*, December 21, 2002.

Websites
The Key to Everything, from the University of Michigan
http://www.medicineatmichigan.org/magazine/2002/winter/stemcell/default.asp

Sources of Stem Cells, from the National Marrow Donor Program
http://www.marrow.org/MEDICAL/sources_of_stem_cells.html

The Stem Cell Debate, from Time, Inc. http://www.time.com/time/2001/stemcells/

The Stem Cell Debate, from CNN http://www.cnn.com/SPECIALS/2001/stemcell/

Stem Cells: A Primer, from the National Institutes of Health
http://www.nih.gov/news/stemcell/primer.htm

Stem Cells: Scientific Progress and Future Research Directions, from the U.S. Department of Health and Human Services
http://www.nih.gov/news/stemcell/scireport.htm

"Under the Microscope: Looking at Stem Cell Research," the National Academy of Sciences
http://www4.nas.edu/onpi/webextra.nsf/44bf87db309563a0852566f2006d63bb/b7ea37ed3bcd3c5c85256ac20076452b?OpenDocument

ABOUT THE AUTHOR

Jenny Tesar has written more than three dozen books for students and the general public. Most of the books, including many titles for Blackbirch Press, have covered the sciences. With Bryan Bunch, she coauthored *The Blackbirch Encyclopedia of Science & Invention* and *The Penguin Desk Encyclopedia of Science and Mathematics.*

INDEX